# Organic Production of Coffee, Okra, Tomato, Mango, and Banana

Gowri Vijayan

All Rights Reserved. No parts of this publication may be reproduced, stored in a retrieval system, or transmitted, in any form or by any means, electronic, mechanical, photocopying, recording, or otherwise, without the prior permission of agrihortico

© 2014 AGRIHORTICO

## Preface

This small book is a sincere attempt to provide some basic insights into organic crop production practices and organic certification procedures, especially for those who have no idea about the topic of organic farming. Techniques mentioned in this book are strictly in an Indian context. Production techniques and all other information had been written keeping in mind the present scenario of Indian organic market and crop production practices. However, this book may be used as a reference material for all situations pertaining to organic production technology.

# TABLE OF CONTENTS

## COFFEE (*COFFEA* SP.) ......................................................... 9

- Introduction ........................................................................ 9
- Selection of site ................................................................. 9
- Varieties ............................................................................. 9
- Raising a nursery ............................................................... 9
- Land preparation ............................................................... 9
- Soil conservation ............................................................. 10
- Preparation for planting ................................................. 10
  - *Shade trees* ................................................................... 10
  - *Green manuring* ............................................................ 10
- Weed management .......................................................... 11
- Intercropping ................................................................... 11
- Training and pruning ..................................................... 11
- Handling, centering and desuckering ......................... 12
- Pest management ............................................................ 12
  - *White stem borer (Xylotrechus quadripes)* ................. 12
  - *Coffee berry borer (Hypothenemus hampei)* ............... 12
  - *Shot hole borer (Xylosandrus compactus)* ................... 13
  - *Mealy bugs (Planococcus citri; P. lilacinus)* ............... 13
  - *Root lesion nematode (Pratylenchus coffeae)* .............. 14
- Disease management ..................................................... 14
  - *Leaf rust (Hemileia vastatrix)* ..................................... 14
  - *Black rot (Koleroga noxia)* ........................................... 14
  - *Coffee trunk canker (Ceratocystis fimbriata)* .............. 14
- Harvesting ........................................................................ 15

## OKRA (ABELMOSCHUS ESCULENTUS) ..................... 16

- Introduction ...................................................................... 16
- Climate .............................................................................. 16
- Soil ..................................................................................... 16
- Intercropping and crop rotation ................................... 16
- Crop duration ................................................................... 16

| | |
|---|---|
| Varieties | 17 |
| Seed selection and treatment | 17 |
| Field preparation | 17 |
| Sowing and spacing | 17 |
| Intercultural operations | 17 |
| Weed management | 18 |
| Irrigation management | 18 |
| Manuring | 18 |
| Pest management | 18 |
| *Shoot and fruit borer (Earias vittella, E. insulana)* | *18* |
| *Leaf hoppers (Amrasca biguttula biguttula)* | *18* |
| *Fruit borers (Helicoverpa armigera, Spodoptera litura)* | *18* |
| Disease management | 19 |
| *Yellow vein mosaic/ Vein clearing (Vector: White fly-Bemisia tabaci)* | *19* |
| *Cercospora leaf spot (Cercospora abelmoschi, C. malayensis, C. hibisci, C. hibiscina)* | *19* |
| *Root knot nematode (Meloidogyne incognita, M. javanica)* | *19* |
| Harvesting | 20 |
| Yield | 20 |

## TOMATO (LYCOPERSICON ESCULENTUM) .......................21

| | |
|---|---|
| Introduction | 21 |
| Climate and soil | 21 |
| Crop rotation | 21 |
| Buffer zone | 21 |
| Land preparation | 21 |
| Planting material | 21 |
| Varieties | 22 |
| Seed treatment | 22 |
| Nursery development | 22 |
| Planting | 23 |
| Irrigation | 23 |
| Cultural practices | 23 |
| Manuring | 23 |
| Pest management | 23 |
| *Fruit borer (Helicoverpa armigera)* | *24* |

 *Serpentine leaf miner (Liriomyza trifolii)*..................................................24
 *Tobacco caterpillar (Spodoptera litura)*.................................................25
 *White fly (Bemisia tabaci)*.....................................................................25
DISEASE MANAGEMENT ..............................................................................26
 *Damping off (Pythium aphanidermatum)*..............................................26
 *Early blight (Alternaria solani)* ............................................................26
 *Fusarium wilt (Fusarium oxysporum f.lycopersici)*................................27
 *Powdery mildew (Leveillula taurica and Erysiphe polygoni)*.....................27
 *Bacterial canker (Clavibacter michiganense)* .........................................27
 *Bacterial wilt (Pseudomonas solanacearum)*...........................................27
 *Leaf curl (Gemini virus)*.......................................................................27
HARVESTING ..............................................................................................28
YIELD ........................................................................................................28

## MANGO (MANGIFERA INDICA) ..............................................................29

INTRODUCTION............................................................................................29
CLIMATE AND SOIL......................................................................................29
VARIETIES ...................................................................................................29
PROPAGATION .............................................................................................29
PLANTING ...................................................................................................30
TRAINING AND PRUNING .............................................................................30
FERTILIZER APPLICATION ...........................................................................30
CULTURAL ACTIVITIES ................................................................................31
PEST MANAGEMENT ....................................................................................31
DISEASE MANAGEMENT ..............................................................................31
HARVEST AND YIELD...................................................................................32

## BANANA (*MUSA* SP.) ..................................................................................33

INTRODUCTION............................................................................................33
CLIMATE AND SOIL......................................................................................33
LAND PREPARATION....................................................................................33
VARIETIES ...................................................................................................33
PLANTING MATERIAL ..................................................................................33
PLANTING AND INFILLING ..........................................................................34
INTERCROPPING..........................................................................................34
MANURING .................................................................................................34

| | |
|---|---|
| *Green manure* | *34* |
| IRRIGATION | 34 |
| INTERCULTURAL ACTIVITIES | 35 |
| PEST MANAGEMENT | 35 |
| *Banana pseudostem weevil (Odoiporus longicollis)* | *35* |
| *Banana rhizome weevil (Cosmopolites sordidus)* | *35* |
| *Nematodes* | *36* |
| DISEASE MANAGEMENT | 36 |
| *Sigatoka leaf spot (Mycosphaerella sp.)* | *36* |
| *Panama wilt (Fusarium oxysporum f.sp.cubense)* | *36* |
| *Bunchy top disease* | *36* |
| HARVESTING | 37 |
| YIELD | 37 |

# Coffee (*Coffea* sp.)

## Introduction
Organic coffee is produced by over 20 countries including Mexico, Bolivia, Costa Rica, India, Madagascar, Brazil, Vietnam etc. Japan, USA, EU are some of the biggest consumers of organic coffee in the world. In India, coffee is mainly cultivated in states of Karnataka, Kerala and Tamil Nadu. India is the only country wherein *Coffea arabica* and *Coffea robusta* coffee are cultivated in almost equal proportions.

## Selection of site
Features like altitude, wind velocity, rainfall, land slope etc. have to be considered while selecting site for coffee plantation. While Arabica coffee prefers higher altitude (1000-1500 above MSL), robusta coffee is better suited to lower altitudes (500-100 above MSL). Sufficient shade should be available for the coffee plants. In case of windy regions, wind breakers like silver oak could be planted.

## Varieties
Hardy varieties suitable for local conditions should be selected. Arabica varieties like S.795, Sln.5-B and robusta varieties like S.274 and CxR are good choices.

## Raising a nursery
Seeds or organic origin is to be used. However, in the absence of organic source, seeds from conventional estates not chemically treated could be used. The nursery beds for organic crop should be separated from conventional nursery, in case the estate is not fully converted.

## Land preparation
Presence of evergreen trees to filter shade is always desirable for coffee cultivation. The spacing of the trees should be maintained at 9-12 m for desirable effects. The land should be planned for division with sufficient footpaths and roads in the middle. Any

bushy growth should be cleared off the field. The land should be tilled and ready before the oncoming rains.

## Soil conservation

The use of two tier shading system, with lower canopy of trees like dadap (*Erythrina lithosperma*) and top canopy of permanent trees, reduce heavy loss of soil by erosion. Depending on the slope of the land, conservation measures like contour and terrace planting are followed.

## Preparation for planting

The spacing recommendation for Arabica coffee is 1.8m x 1.8 m / 2.1m x 2.1 m / 1.8m x 2.1 m and for robusta coffee is 3m x 3 m. Land preparation should be immediately followed by planning on position and spacing of shade trees and coffee plants. Soil exposure to sunlight during the hottest months of the year (April-May) is recommended for killing any soil borne pathogens and pests (root grubs, nematodes etc.). This requires opening up pits of 45cm x 45cm x 45 cm dimensions to sunlight for a fortnight, then filled up with top soil and FYM or compost (1-2 kg/pit). Application of neem cake (250g/pit) is advised before planting, as precaution against root grubs.

### Shade trees

Lower shade canopy trees (e.g. dadap) are recommended at close spacing (4.5-6m apart), for protection of younger coffee plants. Upper canopy trees, of permanent nature (*Artocarpus, Ficus* sp.) need to be planted at spacing of 9-12 m. The soil sterilization process should be carried out before planting of the shade trees. The optimal shade requirement is 50% for Arabica coffee and 30% for robusta. Optimal maintenance of shade will help bring down incidences of white stem borer, green scale, leaf rust, black rot etc. in arabica coffee and shot hole borer in robusta.

### Green manuring

Cultivation of cow pea and horse gram for 2-3 years before coffee helps to build up soil fertility of new farms. With contribution of nitrogen, the green manure crops would also control weed growth in the field.

## Weed management

Weed control is a major problem in coffee plantations, especially new clearings. Grasses have to be weeded continuously in the initial years. Cover digging (30 cm) during planting year and scuffling (15 cm) for the first 3 years post monsoon, helps in weed management and soil moisture conservation in levelled fields. However, sloppy terrain permits only slash weeding, due to high soil erosion tendencies. In the earlier stages of coffee growth, mulching and use of green and cover crops help weed control.

## Intercropping

Intercropping of coffee with short duration fruits and vegetables like ginger, yam, pineapple, banana and papaya have been practiced in India. One of the most common practices is coffee cultivation with pepper. Vanilla is also seen as a good alternative.

## Training and pruning

Training of the plants into bushes will help the crop production processes like spraying and harvesting. Shaded coffee is recommended single stem system of training, with topping at prescribed heights. Arabica (tall) gets topped twice, while arabica (dwarf) and robusta are topped once. The recommended heights of topping are given below:

Arabica (Tall): Topping at 0.75 m

                Topping at 1.35-1.5 m (after 4-5 harvests)

Arabica (dwarf): Topping at 0.9-1.5 m

Robusta: Topping at 1.35-1.5 m (slant cut)

Single stem trained plants require regular pruning (once a year), after harvest of the crop, especially arabica. While pruning, care should be taken to avoid cutting off primary branches to their base, since it is non-regenerative. Instead, some of the nodes at the base of the primary branches have to be left. Removal of secondary and tertiary branches is necessary to encourage new branching. All the suckers and diseased parts have to be removed.

Robusta is a self-pruning crop, which sheds its laterals after 3-4 harvests. Therefore, regular pruning of the plant is not necessary. However, removal of any diseased parts and suckers should be followed 3-4 times a year.

## Handling, centering and desuckering

The thinning off young growth after main pruning is referred to as handling. This should be practices 1-2 times a year, depending of crop growth characteristics. The first handling is done during onset of monsoon season (June-July) and if necessary, followed up in September. Removal of young flush from the primary stem, retaining 4-6 healthy secondaries is necessary. All new growth on the main branches within radius distance of the main stem need to be removed. This is called centering. The suckers have to be removed all the time, 3-4 times a year.

## Pest management

### White stem borer (Xylotrechus quadripes)

-Optimal shade management

-Sanitation and sterilization of field before planting

-Collar pruning of infected plants

-Phyto-sanitary measures

-Removal of any loose barks and thick coffee leaves for egg laying by the pest

-Neem oil spray (2-5%) on main stem fortnightly

-Application of 10% lime to main stem and thick primaries

### Coffee berry borer (*Hypothenemus hampei*)

-Optimal shade

-Good drainage system

-Harvest at proper stage

-Complete harvest

-Reduce gleaning by spreading gunny bag/polythene sheets on the ground while harvesting

-Drying of coffee to prescribed moisture levels

-Phyto-sanitation

-Use of oil smeared polythene sheets to cover harvested fruit heap, to trap the beetles

-Use of trap crops around drying yard

-Post-monsoon spray of *Beauveria bassiana*, an entomopathogenic fungus

-Release of parasitoid *Cephalonomia stephanoderis* in field, post-harvest to reduce any inoculation over crop residues

-Fumigation of coffee during storage, with permitted chemicals

-Placement of berry borer traps, post- harvest to trap the adult beetles

### Shot hole borer (Xylosandrus compactus)
-Shade management

-Drainage

-Frequent pruning

-Phyto-sanitation

-Desuckering

### Mealy bugs (Planococcus citri; P. lilacinus)
-Release of parasite *Leptomastix dactylopii*

-Shade management

-Drenching root zone area with neem oil solution

-Neem oil (3%) spray is effective against mealy bugs and other sucking pests

### Root lesion nematode (*Pratylenchus coffeae*)
-Phyto-sanitation

-Soil sterilization every year

-Use of tolerant rootstocks (robusta, excels, arnoldiana) for grafting of Arabica

-Neem cake application (250 g/plant)

## Disease management

### Leaf rust (Hemileia vastatrix)
-Optimal shade management

-Regular pruning

-Bordeaux mixture (0.5%) pre-monsoon, mid-monsoon and post-monsoon spray

-Spray of Bordeaux mixture during pre-blossom period, in die-back affected areas.

### Black rot (Koleroga noxia)
-Thinning out of shade

-Frequent pruning

-Centering and handling

-prophylactic sprays of Bordeaux mixture (1%)

### Coffee trunk canker (*Ceratocystis fimbriata*)
-Shade management

-Phyto-sanitation

-Collar pruning of infected plants and cut surface treatment with Bordeaux mixture

-Prevent injury to stem

## Harvesting

Harvesting of coffee is done as per the requirements of the produce. The wet method of processing is used for arabica coffee while the dry method is used for robusta. The standards restrictions for organic produce are to be remembered while processing of the produce, especially in presence of non-organic produce. No chemical method can be used for processing of coffee.

# Okra (Abelmoschus esculentus)

## Introduction
Okra belongs to the Malvaceae family and is an annual crop of the tropical and sub-tropical regions. The fruits are slimy, mucilaginous and rich in iodine content which helps in goiter control. Indian states of Uttar Pradesh, Bihar, Orissa, West Bengal, Andhra Pradesh, Maharashtra and Gujarat lead in okra production in the country.

## Climate
Okra is not altitude sensitive but require warm, humid growing conditions. Seed germination requires minimal temperature of 20°C. . The crop is frost sensitive and not suitable for areas with prolonged cold spell. A temperature range of 24°-28°C is most suitable for overall crop growth. Higher temperatures boost vegetative growth rather than reproductive growth of the crop. Temperature crossing 42°C results in flower drop and thereby yield loss in plants.

## Soil
Okra is especially partial to well drained, light and loose loamy soils. Okra can be grown in mild saline soils also. Manuring of soil before crop production is needed.

## Intercropping and crop rotation
Okra is known to respond well in intercropping conditions. Some of the successful intercropping combinations result from combining okra with French beans or radish. The pattern of okra-cowpea-maize, maize-okra-radish and okra-okra-radish have been recorded to reduce the incidence of bacterial wilt in tomato and brinjal, when grown as succeeding crops. Sugarcane yield increase has been reported through intercropping with okra.

## Crop duration
The total crop duration is 3-4 months, depending on variety.

## Varieties

Pusa Sawani, Pharbhani Kranti, Janardhan, Anamika, Varsha Uphar etc. are some of the popular okra varieties. Selection of variety for organic farming should be done based on its disease and pest resistance among other specifications.

## Seed selection and treatment

Seeds can be procured from a certified organic farm or from self-plot. In case of self-plot, care should be taken to isolate the plants from contamination and selection of disease and pest free sources. Once the pods are mature, dry and start cracking, they can be harvested. In order to sow in 1 hectare of field, about 300 plants are to be isolated. A seed rate of 22 kg/ha for a summer crop and 12 kg/ha for a rain-fed crop is optimal. Seed treatment is essential not only as a preventive measure, but also will aid in germination of seeds. In case of summer cultivation, seeds should be soaked in water for 12 hours prior to sowing. Seed treatment with sweet flag rhizome extract for 30 minutes prior to sowing gives resistance to fungal and bacterial diseases. Seeds can also be treated with jeevamrut for 4-6 hours after soaking in water for 8 hours, shade dried and sown.

## Field preparation

Proper tillage of the land to a fine tilth is required for okra plant.

## Sowing and spacing

Okra seed sowing is done either by dibbling or using a seed drill or plough. Broadcasting of okra seeds is not common. In order to ensure proper drainage, reduced water requirement and good germination, ridge sowing is adopted. Seeds are dibbled at the rate of 2-3 per hole. The spacing specification changes with varieties, however ridge to ridge spacing is recommended at 45 cm. The general spacing for branching varieties if 60cm x 30 cm and for hybrids is 75cm x 30 cm.

## Intercultural operations

Once the germinated seedlings attain 1 true leaf stage, thinning out of closely germinated seedlings is to be done.

## Weed management

Hoeing and weeding need to be practiced regularly as per requirement. The standard weeding schedule is 20 days after sowing, then every 25$^{th}$ day after. Earthing up is done 1 months after sowing.

## Irrigation management

Soil moisture is essential for seed germination. In case of less moisture, watering of the ridges should be done before sowing. In case of high temperature, light irrigations should be provided for proper fruiting. Water logging should be avoided. Drip irrigation is not commercially used for okra and furrow system is commonly practiced. Flowering and fruit set stages of okra requires sufficient water, otherwise moisture stress will lead nearly 70% crop loss.

## Manuring

Yearly application of FYM to the soil increases the yield of the crop. Application of FYM at 25t/ha, neem cake at 250kg/ha and groundnut cake at 80 kg/ha are recommended. Addition of biofertilizers to the soil is also advised.

## Pest management

### Shoot and fruit borer (Earias vittella, E. insulana)
-Early sowing of okra seeds to avoid damage during rains

-Soil sterilization

-Summer ploughing

-Spraying ginger, garlic, chili extract

### Leaf hoppers (Amrasca biguttula biguttula)
-Installation of yellow sticky traps (30/ha)

-Spraying 5% need seed kernel extract

### Fruit borers (Helicoverpa armigera, Spodoptera litura)
-Summer ploughing

-Installation of bird perches and use of bird attractants to attract predatory birds

-Fenugreek flour (1 kg) + 2liters of water. Mix them and keep aside for 24 hours. Afterwards, dilute with 40 liters of water and spray on crops (solution for 1 hectare).

- Placing pheromone traps (8/ha)

-Use *Trichogramm*a (50,000 eggs/ha) 6 times every alternate week

## Disease management

### Yellow vein mosaic/ Vein clearing (Vector: White fly-*Bemisia tabaci*)

-Cut a milk bush (*Euphorbia tirucalli*) or cactus (e.g. *Euphorbia nivulia*) into pieces, immerse in water and allow fermentation for 2 weeks, after which filter the solution and spray on the crop

-Vector control using yellow sticky traps

-Weed hosts destruction in the vicinity of the crop

-Avoid planting during hot season

-Use of resistant varieties like Arka Anamika, Pusa A-4 etc.

-Phyto-sanitation

-Spray of 5% neem seed kernel extract

### Cercospora leaf spot (Cercospora abelmoschi, C. malayensis, C. hibisci, C. hibiscina)

-Spray of 5% neem seed kernel extract

-Cleaning of bunds and phyto-sanitation

### Root knot nematode (Meloidogyne incognita, M. javanica)

-Practice of crop rotation with cereals

-Use of marigold as an intercrop

-Addition of neem cake (25qtl/ha) to soil

-Use of bio-control agents *Paecilomyces lilacinus*, *Bacillus penetrans*

Some general management practices include:

-Mint leaf extract (250 g powder/2 liters of water) spray

-5% neem seed kernel extract spray

-Fumigation with *Embelia ribes* or *Acorus calamus* during evening hours

## Harvesting

The harvesting period depends on the variety characteristics and ranges from 45-60 days after sowing. The pod size and stage depends on the market preference. Harvesting is normally conducted during morning hours. The variation in maturity of fruits of the same crop influences frequent harvesting. This also helps regulate pod size, maturity and enhances fruit formation. The harvest from mornings immediately go to evening markets .The post-harvest handling of okra for long distance markets require cold chain facility, making it delicate issue for small farmers.

## Yield

Okra generally yields 7.5-10 t/ha on an average, while hybrid varieties reap 15-22 t/ha.

# Tomato (Lycopersicon esculentum)

## Introduction
Tomato is the world's largest vegetable crop after potato and sweet potato. It belongs to the *Solaneceous* family, along with tobacco, potato and bell pepper. Native to Peruvian and Mexican region, it is considered very important due to its nutritive content. The wide versatility in its usage makes it a favorite among food processors.

## Climate and soil
Tomato crop require low-medium rainfall conditions for growth. Winter growing has been found ideal for organic tomato cultivation. Well drained sandy loam soils are best suited for cultivation. Acidity of the soil has to be checked regularly at the start of each season. Extremely acidic soils are to be avoided or corrected for pH level.

## Crop rotation
Crop rotation with non-solanaceous crops like legumes, pulses are recommended. Cropping systems like okra-tomato, tomato-onion is quite popular in India. Crops like rice, cauliflower, watermelon, garlic, cotton, sunflower etc. can also be grown after tomato. A gap of 1 year should be maintained between planting of solaneceous crops (brinjal, capsicum, chili, potato, tomato etc.).

## Buffer zone
Buffer zone of 7.5-15m is required, if non-organic cultivation system is practices by neighboring farms/fields.

## Land preparation
The field should be ploughed to reach a fine texture, after which 10 tons of FYM or vermicompost at the rate of 1-1.5 t/acre should be added and mixed well. Beds are raised after the mixing of manure. Drainage channels are made at 50 cm breadth for every 1 m of the bed.

## Planting material
Tomato plants are generally propagated through seeds. Organic cultivation required the use of certified organic seeds. The selected

seeds should also be of disease resistant, high yielding varieties with tolerance to pests. The seeds can either be purchased from certified organic farms or from own seed plot.

## Varieties

For organic farming of tomato, the open pollinated varieties are preferred. Some of the varieties developed like Swarna Lalima and Swarna Naveen are suitable for organic cultivation. Lakshmi NP 5005 is another popular variety with resistance to bacterial wilt and leaf mosaic virus. Some of the varieties resistance to

-Bacterial Wilt: Arka Abha, Swarna Lalima, Arka Alok, Arka Abhijit (hybrid)

-Leaf curl: Anmol, Aditi (hybrid), Kashi Amrut, Avinash-2 (hybrid)

-Blight: Akash, Vajra, Meghana

-Verticillium and fusarium wilt: Empire, Rupali (hybrid), Roma

-Nematodes: SL-120, Ronita

Varieties with increased shelf life include hybrids like MTH-6, Rupali, Avinash-2, Pusa Ruby, Pusa Early Dwarf, Arka Vikas, Arka Abha, ArkaAlok, Punjab Chuhara, Co 3 etc.

## Seed treatment

Treatment of seeds with Trichoderma at the rate of 1g/150 g of seeds, prior to sowing builds up its resistance.

## Nursery development

Nursery beds of 1m x 3m specification at 20 cm height are to be made. It has been calculated that 12-15 such beds (150 g seeds) suffice for procuring saplings for planting in 1 acre. Standard application of manure to the soil, followed by sterilization by solarization reduces pest incidence during early stages. The seeds need to be planted in intervals of 5 cm with 2 cm spacing between successive seeds planted. The depth of planting should not exceed 1 cm. The covering of beds after sowing, with nylon sheets prevent pest infestation in saplings.

## Planting

Seedlings of 20-25 days can be used for transplanting. The spacing followed for transplanting in main field is 60 cm between rows and 50 cm between plants. The planting should not be too dense to prevent further growth of plants, nor too light for weeds to grow. The irrigation and drainage channels must be distanced from plants by 30 cm.

## Irrigation

Irrigation depends on the soil type, climatic conditions. Generally, the field need to be irrigated once in 7 days.

## Cultural practices

Weed control is an important cultural activity to follow particularly 4-5 weeks after transplanting. Use of mulches, crop rotation, sanitation, shallow tilling etc. need to be followed to prevent growth and spread of weeds. Regular weeding is necessary particularly when the seedlings are small. The tomato plants grow well with staking in the form of small bamboo sticks, branches or wires for support. This should be done 15-20 days after the transplantation or when the seedlings attain 20 cm height. This will help with their branching and increase production.

## Manuring

10t of FYM or 1-1.5 t of vermicompost per acre is the standard application rate. Treatment of FYM with Trichoderma (500g/tractor load of manure) and its application after culturing is highly beneficial and recommended.

## Pest management

Crop rotation is one of the traditional methods used to break the life cycle of pests and insects. Careful monitoring of the field and crop is essential to identify the presence of pests. It has been observed that crop rotation is fairly successful in pest control for organic tomato. Trap crops are also used for preventive control. Sweet corn is a trap crop used to attract the tomato fruit worm. Marigold is used to attract the tomato fruit borer and reduce loss to main crop. For every 16 rows of tomato, one row of marigold should be sown as a trap crop. The marigold seedlings must be 15 days older to the tomato seedlings. This is to facilitate synchronized flowering of the crops. Biological control of pests has been found to be the most effective pest management strategies in case of organic tomato farming. The use of *Trichogramma* is effective against *Lepidopteran* pests of tomato. Neem based pesticides are effective against the fruit borer of tomato. Trichoderma is suitable for seed treatment and effective against root fungi. Neem seed extract and cow dung mixture is also used to prevent excessive flower drop in the plants. Some of the major pests to tomato and their management are given below:

### Fruit borer (Helicoverpa armigera)

-Look out for *Helicoverpa* eggs on the top leaves and hand pick the larvae

-Use of resistant varieties like Rupali, Roma etc.

-Usage of American marigold as a trap crop

-Border cropping using sorghum (8 rows) at 30cm x 10 cm spacing support natural predators like *Coccinellids* and *Chrysoperla*.

-Attracting predatory birds using bird perches or any other luring techniques

-Need seed kernel 5% spray to stop the pest at its early stages

-Severe borer attack can be controlled by application of seed extracts of *Strychnosnux-vomica* to the soil, at the rate of 1.5 g/plant every 20 days (twice)

-Use of biocontrol agents *like Bacillus thuringiensis* (1g/liter of water), *Trichogramma chilonis* (50,000 eggs, six times at weekly intervals), NPV (250 LE/ha) etc.

-Setting up of pheromone traps (15/ha) with Helilure, changing every fortnight

### Serpentine leaf miner (*Liriomyzatrifolii*)

-Intercropping with field beans (1 row for every 8 rows of tomato), with fields beans sown 12 days before tomato transplanting

-Phyto-sanitation

-Need seed kernel extract 5% spray or ginger, garlic, chili extract (1lt/tank)

### Tobacco caterpillar (*Spodoptera litura*)

-Exposure of soil to extreme heat through ploughing (before transplanting)

-Flooding of field while field preparation, in order to kill hibernating larvae

-Use of castor as trap crop (125/ha) will attract egg laying moths. Eggs of the pest can be collected from castor and destroyed

-Laying pheromone traps (15/ha) with pheroclin SL lure for monitoring pest population

-Neem seed kernel 5% spray for protection against larvae

### White fly (Bemisia tabaci)

-Use of nylon net in nursery to protect the seedlings

-Phyto-sanitation

-Irrigation control

-Use of *Brumus* and *Chrysoperla*

-Use of pearl millet as a barrier crop around main field, sown 15 days prior to main crop transplanting

-Removal of weed hosts

-Yellow sticky traps (50/ha)

-Neem seed kernel extract 5% spray

A general remedy for pest management is the spraying of decoction made of leaves from aloe vera, neem, *Ocimumtenuiflorum*, *Achyranthesaspera* and *Aristolochiabracteata*. The boiled decoction is mixed with water (100 ml/liter of water) and sprayed on the plants.

## Disease management

Tomato faces diseases caused by fungi, bacteria, physiological disorders caused by abiotic/ environmental stress like cat face and blossom end rot. The major diseases that attack the root system of the crop include fusarium wilt, verticillium wilt, bacterial wilt, rhizoctonia. Diseases like early blight, leaf spot, bacterial canker and late blight attack above-ground stems and foliage and bacterial spot, bacterial speck and anthracnose affect the tomato fruits. Pest life cycle con be discontinued if solaneceous crops are not subsequently planted while in rotation.

### Damping off (Pythium aphanidermatum)
-Use of certified seeds

-Sterilization of soil by solarization

-Better drainage facilities

-Neem cake (400 g/sq. m) application to nursery bed 15 days prior to sowing and irrigation at 4 days interval

-Use of light, draining soil for nursery beds, light and frequent irrigation

-Use of well decomposed manure

-Seed treatment with leaf extract of *Bougainvillea glabra* (20ml/liter of water) for 6 hours

### Early blight (Alternaria solani)
-Selective crop rotation

-Distancing tomato and potato cultivation areas

-Sanitation of crop by removing and burning of infested branches, leaves

-Spray of 5% eucalyptus or lantana leaf extract during evening hours

-Treatment with *Trichoderma viride* or *Pseudomonas fluorescens* (5g/100g of seeds)

**Fusarium wilt (Fusarium oxysporum f.lycopersici)**
-Selective crop rotation

-Seedling root tip treatment in solution of turmeric and asafetida (10 g each/liter of water) prior to transplanting

-Sanitation of field

-Use of resistant varieties

-Spraying of 15 days old diluted panchakavya (1:10)

**Powdery mildew (Leveillula taurica and Erysiphe polygoni)**
-Recommended spray of solution of milk and water (1:1) once in 3 days after disease notification

**Bacterial canker (Clavibacter michiganense)**
-Crop rotation avoiding subsequent rotation of solanaceous crops

-Spraying cow dung extract

-Hot water seed treatment

**Bacterial wilt (Pseudomonas solanacearum)**
-Crop rotation with field beans, maize, soybean or cruciferous vegetables

-Seedling treatment (root dip) in asafetida solution (10g/liter of water)

**Leaf curl (Gemini virus)**
-Crop rotation

-Use of healthy seedlings

-Soil sterilization

-Crop sanitation

-Spray of 5% Neem seed kernel extract to control white flies

One general pest preventive measure taken up is crop rotation with corn, cereals and sorghum. Field sanitation and roughing of affected plants is also essential. Covering of exposed fruits with straw will help prevent sun scald (blotches and dry skin in fruits).

## Harvesting

The crop takes 2-3 months for reaching maturity. Harvesting stage is determined by market requirements. The stages of harvest are immature green, mature green, turning pink, half ripe, red ripe and over ripe. Turning pink fruits just show color at the blossom end while mature green are not pink. Half ripe fruits are majorly covered in pink color. The mature green fruits are best suited for export purposes. For domestic fresh consumption, the turning pink or half ripe stage is the best. For the purpose of seed production, the red ripe tomatoes are best. Canning and processing stages could use the red and over ripe stage fruits as per requirements.

## Yield

Organic tomato yield varies form 15-20 t/acre, under irrigated conditions.

# Mango (Mangifera indica)

## Introduction

Mango is known as the King of fruits. Belonging to the Anacardiaceous family, mango is an evergreen tree. Mango is a very important fruit crop for India. India is the largest producer of mango in the world, responsible for 60% of world mango production. The states of Uttar Pradesh, Andhra Pradesh, Kerala, Tamil Nadu, Karnataka and Bihar are the major producers of mango in the country. Mango is also a major export item for India. Indian mangoes are majorly exported to UAE, Kuwait and other middle-eastern countries. Alphonso and Dashehari varieties are in demand in its fresh form.

## Climate and soil

Mango flowering period is susceptible to high humidity, frost or rains. Therefore, site selection should take climatic parameters into consideration. Dry summer and good rainfall is an ideal situation for mango cultivation. But, areas reporting high winds and cyclones should not be used, since it might result in excessive flower and fruit dropping and damage to branches. Mango can be cultivated in a wide range of soils, provided well drained and deep, still slightly acidic soils (pH 5.5 to 7.5) are preferred.

## Varieties

The selection of varieties for organic farming should be based on the hardiness and pest/disease tolerance of the crop. This is a crucial preventive strategy for pest and disease management in organic farming. There are many varieties developed over the years for various purposes. Some of the most famous varieties are Alphonso, Neelam, Mulgoa, Banganpalli, Dashehari, Kesar, Bombai, Chausa, and Fazli. Some of the mango hybrids developed are Amrapali, Mallika, Mangeera, Ratna, Arka Puneet and Arka Aruna.

## Propagation

Farmers should only procure vegetatively propagated, true to type healthy plants from organically certified nurseries. The plants should be checked for pest or disease attack before buying. Some of the popular methods of propagation in mango are inarching, veneer grafting, side grafting and epicotyl grafting. Stone grafting is the most successful process in mango.

## Planting

Deep ploughing followed by harrowing and levelling of the land should be done. The drainage channels should be well planned and constructed. A spacing of 10m x 10m is recommended for dry zones, while 12m x 12m spacing is recommended for heavy rainfall areas. Planting should be done based on square or hexagonal system of planting. Before planting, pits should be filled with a mixture of top soil and 10 kg FYM/ compost per pit. Evening hours are preferable for planting. The planting should not be done too deep. Planting of the grafts should be done with the graft union above ground level. The plant should be immediately irrigated. The first 1-2 years requires provision of shade for the young plants and staking of the grafts when necessary.

## Training and pruning

The main stem should be allowed to branch out and cross over branches removed when young.

## Fertilizer application

Recommended FYM/compost application schedule for organic mango cultivation is given below:

1$^{st}$ to 2$^{nd}$ year - 15kg/plant

3$^{rd}$ to 5$^{th}$ year – 30 kg/plant

6$^{th}$ to 7$^{th}$ year – 50 kg/plant

8$^{th}$ to 10$^{th}$ year – 75 kg/plant

Oil cake (10kg/plant) and green leaves (25kg/plant) can be added to as additional sources. Trench system should be adopted for application of the manures (2.5m – 3m away from base).

## Cultural activities

Continuous irrigation (twice a week) is necessary until the mango trees reach 4-5 years of age. Intercropping with vegetables, black gram, horse gram, pineapple and banana have been proven fruitful. Intercultural activities should be carried out twice a year. Irrigation schedule should be fortnightly after initiation of fruit setting stage of the crop.

## Pest management

Some of the important pests of mango are hoppers, stem borers, shoot midges, leaf feeding insects, fruit flies and mealy bugs. Some of the common pest management practices followed are:

-Phyto-sanitation of the orchard

- Smearing of lime on tree trunks

-Closing up of open wounds on plants with Bordeaux mixture (1%)

- Smoking of orchards during flowering season, helps to control hoppers

-Spray of *Pseudomonas fluorescens* (10g/lt) before flowering and during flushing helps resistance building in trees

-Spathiphyllum (Peas Lily) can be used as a trap crop

-Burning of rotten and damaged fruits

-Methyl euginol traps (10/ha)

-Spray of neem oil and soap emulsion to control sucking insects

-Spray of 1% soap solution + 1% pure alcohol helps control mealy bugs

## Disease management

The most common diseases are powdery mildew, anthracnose and dieback.

-Phyto-sanitation

-Removal of infected twigs and spray of 1% Bordeaux mixture helps control die back

-Application of 10% Bordeaux paste below infected portion of bark, to control pink disease

-Burning of infected fruits and branches

-Hot water treatment of infected fruits could reduce the fungal growth

-Proper drainage

## Harvest and yield

The harvesting stages of mango depend on its market requirements. Mango plantation becomes viable commercially 4-5 years after planting. The fruit color change from green to red or yellow and changing texture of peel signals maturity for harvest. For long distance markets, mangoes are picked before reaching the extreme ripe stage. Care must be taken to avoid any bruising to the fruits while picking. The harvested fruits are to be treated with hot water followed by 8% plant wax, to reduce anthracnose disease. After this, packing of the fruits is done according to market distance and intention for produce. The yield goes from 30 kg/tree in the 4th year to 100 kg/tree by the 10th year after planting.

# Banana (*Musa* sp.)

## Introduction
Banana is a perennial herb, native to South East Asia. It is the fifth largest agricultural commodity in world trade after cereals, sugar, coffee and cocoa. Ecuador, Brazil, China, India are the leading producers of banana in the world. USA, Belgium, Germany and United Kingdom are the major importers of banana. Some of the top banana producing states of India are Tamil Nadu, Maharashtra, Andhra Pradesh and Karnataka.

## Climate and soil
Banana is a year round crop with temperature requirement of 25°C – 30°C. Deep well drained soils with pH 5.5-7.5 are preferred for banana cultivation. Sandy loam and black loamy soils are the most suited for cultivation.

## Land preparation
Land preparation starts with ploughing (2 times) followed by harrowing (3 times). It is followed by levelling using tractor. Ridges, furrows are made and pits (30cm x 30cm x 30cm) dug up. FYM is filled into pits and irrigated.

## Varieties
Some of the varieties used for organic cultivation include Dwarf Cavendish. Grand Naine, Basrai and Sreemanthi. Njalipoovan, Palayankodan, Robusta, BRS-1 and BRS-2 are suited for intercropping with coconut. Dudhsagar is highly resistant to major pests and diseases. Manjeri Nendran 2, BRS-1, BRS-2 are some of the varieties less susceptible to sigatoka spot disease.

## Planting material
The planting material for banana are suckers and rhizomes. Healthy sword suckers of 3-4 months age should be selected from diseases free clumps. Hot water treatment of the rhizomes is recommended to prevent any nematode infestation.

## Planting and infilling

Planting of banana should not be done in the rainy season. Before planting, planting material treatment in a solution of 250 g ghee + 0.500 g honey + 15 kg cow dung or dip in *Pseudomonas fluorescens* (2%) solution for 30 minutes before planting is considered beneficial. The spacing adopted depends on the variety and area of cultivation. The general spacing adopted is 1.5m x 1.5m and 1.5m x 1.8m. A population of 4400 plants could be accommodated with the spacing of 1.5m x 1.8.

## Intercropping

Banana is generally intercropped with onion, soybean, and cowpea. Along with suppressing weed growth, these intercrops also act as an additional source of income for the farmers. Other profitable intercrops include Amaranth, yam and colocasia. .

## Manuring

A basal dose of FYM (15 t/ha) is to be applied during land preparation. A schedule of 5 kg FYM + 20g Trichoderma + 30-50 ml humus per plant for the first year and 10 kg FYM per plant for the 2$^{nd}$ and 3$^{rd}$ year, can be opted. The spraying of jeevamrut is also a good practice. Foliar spraying of panchakavya 3% at 3$^{rd}$, 6$^{th}$ and 9$^{th}$ month after planting is recommended.

### Green manure

Green manure crops sunhemp, diancha, cowpea can be sown after planting of banana, at the rate of 50kg/ha. The crops should be cut and incorporated into the soil after 40 days of sowing. This can be again repeated.

## Irrigation

The plants should be irrigated after planting. An irrigation schedule of 40 weekly irrigations is necessary. Since inadequate irrigation leads to delayed flowering, irregular bunch size, delayed maturity, reduced finger and reduced quality of the fruits, monitoring of moisture level is necessary. Once in three days irrigation schedule will have to be planned in case of increase in temperature. Water stagnation should be prevented with good drainage system. This will help reduce any pest and disease incidence by damping.

## Intercultural activities

Mulching should be practiced immediately after planting of banana. It not only conserves moisture but also help in weed control. Propping/supporting of the fruit bearing plants need to be done using bamboo or wooden poles. Desuckering every 20 days is to be practiced to prevent competition for nutrients among the suckers. This has to be continued toll flowering stage of the crop. Once formed, the branches have to be wrapped up, to get bruise free and uniform fingers.

## Pest management

### Banana pseudostem weevil (*Odoiporus longicollis*)

-Field sanitation

-Burning up of infected plants

-Pseudostem of harvested plants should be destroyed

-Remove loose sheaths of pseudostem and swab mud slurry + neem oil emulsion (3%) on it

-Trap weevils in split pseudostem trap, collect and destroy daily

### Banana rhizome weevil (*Cosmopolites sordidus*)

-Selection of disease free planting material

-Deep ploughing and exposure of soil to sunlight

-Hand removal of eggs and young ones of weevil

-Trap weevils in split pseudostem trap, collect and destroy daily

-Set up pheromone traps with Cosmolure/ Cosmolure⁺ and change the pheromone sachets every 45 days

-Application of crushed need seeds (1kg/plant) to the pit

**Nematodes**

Root knot nematode (*Meloidogyne incognita*), burrowing nematode (*Rasopholus* sp.), root lesion nematode (*Pratylenchus coffeae*), cyst nematode (*Heterodera oryzocola*)

-Hot water treatment of rhizomes

-Neem cake application (1kg/plant) during planting

-Intercropping with marigold and sunhemp

Aphid (*Pentalonia nigronrevosa*)

-Use of bio-control agent Verticillium lecanii (1 x $10^7$ spores/ml)

## Disease management

### Sigatoka leaf spot (*Mycosphaerella* sp.)
-Burning of affected plants

-Restricted spraying of Bordeaux mixture (1%)

-Use of bio-agents like *Pseudomonas fluorescens* (20g/liter) or *Bacillus subtilis* (5g/liter)

-Use of resistance varieties like Dudhsagar or BRS-1.

### Panama wilt (Fusarium oxysporum f.sp.cubense)
-Removal and burning of affected clumps

-Lime application (500 g/pit)

-Neem cake application (1 kg/pit)

-Treatment of planting material in 2% *Pseudomonas* before planting

-Use of resistant/ less susceptible varieties like Nendran, Robusta

### Bunchy top disease
-Use of disease free suckers

-Phyto-sanitation

-Neem based insecticidal spray

-Use of bio-control agent Verticillium lecanii ($1 \times 10^7$ spores/ml)

- Use of tolerant varieties like Njalipoovan, Karpooravally

## Harvesting

Banana takes nearly 12 months for reaching maturity for harvest. The first ratoon crop would be available for harvest with 10 months of the main crop harvest and the second ratoon at 8 months of first ratoon harvest.

## Yield

Average yield is about 70 t/ha by the 3rd year of cultivation.

www.ingramcontent.com/pod-product-compliance
Lightning Source LLC
Chambersburg PA
CBHW051826170526
45167CB00005B/2170